A 21st Century School

Build a School for Tomorrow and Build a Legacy

By Kent Pilcher

Here's What's Inside...

Introduction

May 2014
Quad-City, Iowa

One of the questions superintendents often ask me is how do I take on the daunting task of advancing my district while still doing my job. They often say they don't know how to begin to think about the steps that will lead to a  successful outcome; building a legacy for the district and community.

This book is the result of that question and is designed to help you think about considerations that will lead to success.

What follows is a framework that will help you develop a 21st Century model to deliver 21st Century learning for you, your board and your community. All during the way keeping in mind, "what's best for the kids." This model will focus on managing the varied and multiple stakeholders and conflicting priorities to create facilities that will advance learning and collaboration allowing the students to better achieve.

I hope this book educates and encourages you to develop your way of thinking about how to go about your next building project. I also hope this motivates you to use a collaborative process and not go it alone by taking advantage of others' expertise.

Kent Pilcher

A 21st Century School!

Susan: Good afternoon, this is Susan Austin and with me today is Kent Pilcher from the Quad-City region of Iowa, welcome Kent.

Kent: Thank you Susan, glad to be here.

Susan: We are going to be talking about your book *A 21st Century School*. Why did you want to write this book, Kent?

Kent: Susan, in our line of work, we see firsthand what happens when a school superintendent is faced with a need to address changes in their school district, whether it's expansion, efficiencies or consolidation of schools, and they don't have any idea where to begin to think about how to best handle this process. As a result, they often times don't think comprehensively about the best process, so they can ensure a successful outcome for everyone. In other words, how do you start and end this process very successfully, so you create a tremendous outcome for your district and for the community.

Susan: You're talking specifically about the superintendents who are responsible for the building of these new schools. Is that correct?

Kent: That's correct. The superintendents are charged with taking the lead on behalf of their school district to begin to address what's necessary to meet the changing needs of their district.

Why This Is Much More than a Building Project...

Susan: What do the superintendents need to understand when they take on a school project for their district?

Kent: First of all, this is much more than just a building project. This is a collaborative process of engaging multiple stakeholders, all of whom have varied and different interests, whether it be the staff and administration within the district, whether it be the board of directors for the school district or whether it be the entire community. The challenge is not about a school project. The challenge is how to develop a collaborative process that brings the community together where the kids really benefit in the end.

Susan: Can you describe for us who the key stakeholders are?

Kent: Certainly. The initial players really are the superintendent, and usually their top staff, who have been thinking about this project and want to really advance their concept or an idea. The next level is the people who are going to be affected by the idea. It might be people out in different buildings. It could be the teachers. It could be the people who are responsible for the curriculum, all within the district, who are in charge of delivery of the educational product. In addition to those people, the superintendent has to work with their board to make sure the board is comfortable with the vision because the board has to then answer to the community.

The board in turn, is responsible for what makes sense financially for the community on behalf of the community. Those are the primary players involved. If you envision an hour glass, the superintendent and their key staff are in the middle. The board is above the middle facing outward toward the community and the staff and the people delivering the education to the kids within the district are below the middle. The "pinch point" is at the middle of the hourglass.

Susan: This is very early in the process right? Have they even determined if they will be building a school at this point?

Kent: Exactly, yes. This is really at the idea or concept stage of the project. There may be some growth in the district and they may have some needs created from future growth they anticipate, or there may be a need due to some shrinking within the district. There may be operational pressure on their budget where they have to look at streamlining things in the district to reduce costs and so they may want to consolidate buildings. They may have aging schools or facilities that are energy inefficient they want to consider replacing. This whole process starts around a question or an idea because the district is changing dramatically and the superintendent wants to get out in front of the impact of those changes. In the sense of these broad questions, often times superintendents or boards don't really even know how to begin to address some of these questions and to understand what their alternatives are.

Susan: When you say the district is changing radically, what do you mean?

Kent: For example, school enrollment, in the upper Midwest is moving from rural to urban and it's creating a lot of different dynamics. Education has become competitive because students now have a choice of open enrollment they didn't have 10 years ago. Now students can enroll anywhere they want. The more desirable schools and districts are the ones that are going to attract or retain their students. That's a different world than it used to be. This is a new challenge for school districts, which plays heavily into the decision making process for the superintendents and boards.

Susan: Interesting. So this isn't a case where the superintendents have a stamp of approval to build a certain school with a certain budget. You're talking about how they even go about deciding what is best for the district in terms of developing a strategy and possible solutions. You want to help the superintendents develop ideas, options and a process on how to move the district forward.

Kent: Exactly. This is really a question of how they explore what all the alternatives are, so there isn't even a decision to be made just yet. They're trying to determine, "What options do we have to solve this? Do we build? Do we expand? Do we shrink? What can we afford? What are all of the options we might have to deliver the best result for the district and make really good informed choices along the way?"

Susan: And the superintendents are the lead on this?

Kent: Yes. They're the CEO of the district and they're charged with providing all the options to the board of education.

Susan: The superintendent has to bring the vision and sell it, if you will, to the school board. Is that correct?

Kent: Yes, or at least to get the board to buy in. It's the superintendent's job to identify options, collaborate with the board to vet the most responsible options and determine a direction.

What Superintendents Need to Understand When Building a 21st Century School...

Susan: Let's talk about some of the things superintendents need to understand when they're taking this on, because I can see that it's not a simple case of designing a pretty building. There's a lot more at stake here.

Kent: Yes, there is. There's the whole question of what are the unmet needs in the district? Maybe we have an old building and we need to refurbish it. Maybe we want 21st century learning environments within the structure of some of our older buildings. How would we accomplish that? It could be any number of questions where they are trying to understand, across the district, what they can do, and then secondarily think of those questions as a sort of a wish list. Here are all of the things I might want to do in the district or might be thinking about doing.

On the other side of the ledger is the ability to pay for the items on the wish list. Oftentimes there's only a certain financial capacity in the district and the ability to assess the capacity and match it up against the wish list, or at least help the board and the superintendent and the staff do that as an essential exercise. Putting together that kind of comprehensive approach and prioritization is really a challenging process for which most superintendents or boards are never trained. The idea is to come up with and address the wish list, determine the costs, assess the financial capacity of the district and then help the district determine, "Can we afford this within our current income stream, or do we need other outside sources like a bond referendum that would provide additional financial capacity?"

Susan: Right, because the superintendents generally come up through a teaching capacity, so managing this project is another level of complexity to figure out.

Kent: Yes. It's a case where aspiring teachers who want to move up in the world become building principals and aspiring building principals become district superintendents. They're really trained in the delivery of education and to some extent, the management of other educators. Where superintendents sometimes need assistance is formulating an approach to develop a concept and then a strategy that will bring together a group of people to align their thinking around one of the best concepts.

How to Accommodate the Delivery of 21st Century Learning...

Susan: Very good, Kent. You titled the book, *A 21st Century School.* I'm curious what some of the differences between a 21st century school and a more traditional school are?

Kent: Twenty-first century schools are a product of how to deliver educational and co-curricular activities for 21st century learning, and it is one of the hottest topics in education right now. It is a question of starting to think about how to align education and learning with how the world has changed and evolved over the last 20 years or so. The delivery of learning and education models needs to be very different today, and certainly more different in the future. Really, if you think back about how education was delivered and schools were built, it was a teacher in a box and the schools were boxes of rooms assembled around other kinds of support areas. Teachers lectured in isolated rooms and students moved from room to room for separate subjects.

The teacher was kind of "the sage on the stage," standing up there presenting education to students, the students individually learned, and the classrooms were individual areas. Now, the evolving concept of 21st century learning and education is based around the idea of collaboration and people working in teams and on projects through interdisciplinary experiences. Twenty-first century learning is built around the idea of collaboration and project-based learning experiences that promote creativity and critical thinking. The delivery is "student centered" as contrasted to "teacher centered." Interestingly, those are skills needed in the workplace today to be successful. As an example, our employment opportunities have dramatically shifted in the last 50 years. If you examine manufacturing, the workforce has dropped from slightly above 25 percent to slightly above 5 percent. On the other hand, in the finance and business services sectors, workforce has risen from slightly below 5 percent to above 15 percent in each of those sectors. Workplaces today require a great deal of collaboration, creativity, critical thinking skills and the ability to be strong communicators. So 21st century learning aligns with the skills needed to be successful in the workplace.

So this presents us with the question, "Okay, now how do we envision these schools to be able to accommodate the delivery of 21st century learning?" Almost all of them were designed and built with a delivery model that strongly contrasts for 21st century. Today, these schools have to be thought of differently, not only today, but to allow for future flexibility. Imagine how fast technology is changing so many things in our world. The schools need to have flexibility for 40 or 50 years into the future in order to be able to continue to adapt based on the speed of the changing technology. These buildings aren't being designed and built for 10 years, but flexible for 50 to 75 years.

Beyond that, in the old days, schools were designed to be open for example at 8:00 a.m. and the day ended at 5:00 p.m., and then they went "dark." Sometimes they wanted another hour or so for athletic, band or theater practices. But today's schools are being thought about in terms of how they interface with the community and not just the students. How they might be 12 to 20-hour per day facilities that can be multipurpose for the community and, more importantly, for the community in lifelong learning? We have to start to think in terms not just inwardly in the delivery of education, but outwardly in terms of how we might promote a learning structure that benefits lifelong learning in the whole community.

How that translates into the actual building becomes a process in how we think about things like the design of the building and what's necessary within the building for these spaces. There are some pretty well-defined concepts in how we think about 21st century buildings where we have much more collaborative spaces that accommodate project based learning. We have libraries, but they're much smaller. Also, today we want to connect students to nature through windows and daylighting. It enhances their learning environment. It is also a very "green" consideration, as well. Those are some of the considerations necessary in school design and construction that will promote 21st century learning. Schools now have to consider things that correspond with the kind of curriculum that is project-based and collaborative-based learning the students are experiencing with the 21st century learning concept.

Susan: The superintendent is the leader or the hub of this process. They are the ones that have to take all these different components and compile them into a vision that will appeal to the board. Some of this is the actual design of the building, but also the larger consideration of what the community might view this effort.

Kent: Right. There's so much "noise," and I'll call it "noise" with quotes around it, out there about what 21st century learning is and what districts should be doing. It's a big buzz right now. The consideration of what it means for how districts teach is something the superintendent has to manage and sell, or develop if it's already gaining momentum. The second consideration is how do we adapt our schools, our physical structures, to that new model? You have parallel paths to consider how education will be delivered and how the schools are being designed and built to facilitate the 21st century methods.

The Tension of Challenging Paradoxes...

Susan: I love that you pointed out that it's not just they're having to design for this next school year that it's going to be open. They have to have the vision to think, "What are the needs of the schools going to be in 20 years, and what do we need to think about now, so we can adapt to that later, without having a huge negative financial impact on the district?"

Kent: That's exactly right. The key here is, we probably can't predict the future very well, but how do we create the most flexibility so the impact of the future changes has the lowest cost? That's the tension in the question. Because, what we do know, for sure, is that Moore's Law of technology says every 18 months the capacity of technology doubles and that's proven itself out since the mid-60s. That acceleration is occurring. Now how that ripples through our society is going to be hard to predict, but certainly we've got to think about creating flexible spaces which are not particularly expensive to change or adapt in order to accommodate things in the future we haven't conceived today.

Susan: Very well said. What else do the superintendents need to consider?

Kent: In addition to 21st century learning, one of the other very hot topics in education right now is how to create safe, secure environments within educational facilities. Obviously, with the number of issues we've seen on national news with people entering schools and doing harm to students, we have this tension in terms of the idea of open environments which are inviting students and the community, but also safe and secure. We have to really think about, "How do we manage that? How do we accomplish that?" Most importantly we have to keep all of that in the forefront with, "What's best for the kids," and the focus on learning for the students. The real question is, how do we create a first-rate, safe, secure 21st Century learning environment, and that applies whether it's a school, an athletic facility, a performing arts center; they're all learning environments.

Susan: The safety factor is, unfortunately, probably one of the most needed items for the districts to address because it's just paramount. Parents want their kids to be safe when they head off to school.

Kent: Yes, as a parent myself, I would want that to be at the forefront of the mind of whoever is in charge of the design. That would be very important.

Susan: What are other challenges these superintendents face when putting together the vision or plan?

Kent: One of the biggest challenges for the superintendent is there's no shortage of what we call resident experts, or strong-willed personalities who might be seen as opinion leaders in a district. Everyone has opinions, whether they're staff, whether they're the board, or they're community people. Education and children are hot topics. Many times, somebody has had some experience, either in the educational field or in the design and construction field, that they feel makes them an expert, and they want to express their opinion.

The challenge, first and foremost, is how do we manage all of these opinions about what we should do, and bring them together to form one vision about what we need to do and where we need to go? Often, what derails community efforts or district efforts is the consideration to get the input from people so that they feel they were heard. For districts to be able to demonstrate that people were heard, and what may or may not be feasible, is extremely important in order to gain support. It is important to really work through that process. Usually, when that does not happen, that's where the proverbial bombshells are discovered.

The whole key here is the need to be organized around a process that's focused on the needs of the district and improving outcomes and learning. That's the ultimate question, because that benefits the students. What's in it for the kids, so to speak? That must be the guiding principle for the superintendent, for the board and for the staff. "How can our kids better learn? How can we create better environments? How can we improve outcomes?" Because improving outcomes is, of course, one of the hottest topics in the country and it was all magnified by the *No Child Left Behind* movement. Now we're into the second and sometimes third generation of what districts are doing about that. The real key here is how do we focus our efforts or our concept around the idea of improving outcomes? In that regard it is not just building a building or buildings, it's improving the outcomes.

How Managing Conflicting Priorities Is Key to Success...

Susan: Yes, I can see this goes way beyond the actual design and construction. There's a lot more the superintendents have to get out in front long before the design or construction. I'm sure the needs and wants of a district and community are large.

Kent: The dynamic here is there are always conflicting priorities, in other words, the needs or the wants of the district are always bigger than the ability to provide funding. So at some point there are competing priorities. Any time you have a situation with competing priorities and varied and multiple stakeholders, you have a challenging situation with potential for conflict. That's what can often happen in these districts as we begin to discuss how they're going to change and evolve. It's really important that we have a collaborative process and educate all of these multiple stakeholders on what the options are and why. It's as much about the process of bringing people together so they understand why not all of the priorities can be met or maybe why the priorities can be met, why the priorities which are being chosen have been chosen and why it's best for the district.

Susan: Is it ultimately up to the superintendent to make those calls?

Kent: I think it's up to the superintendent to help navigate the process with the board, but really the board and the superintendent together make those calls. They should be aligned. The danger is, of course, when the board is not aligned with the superintendent, or the superintendent is not aligned with the board or either is aligned with the community. It ought to be one voice thoughtfully developed by listening and collaborating with multiple stakeholders to get there.

Susan: Is that the reality of what goes on out in the field?

Kent: It's challenging. It's likely to be the biggest challenge these superintendents have to face. Yes, when things come together well, it's like a finely conducted orchestra, so to speak, with one conductor. And the flip side to that is when they don't come together, it's like the proverbial 5th-grade band with no conductor, there are a lot of horns and noise, but not much music.

Susan: What happens Kent, when things don't go well with building these schools? When the players aren't all aligned with the vision?

Kent: Yes, if they don't lose their jobs, they lose their effectiveness with the board, or the district loses its effectiveness as its brand diminishes out in the community. Oftentimes, the path of the superintendent is they start in a small district and they work their way up to a larger district over two or three career moves. One bad experience in a district makes it very difficult to move on to a bigger district. They may not lose their job, but they may have trouble moving on or they may be less effective with their district. It has a huge impact because these kinds of efforts have a lot of intense spotlights from the community.

A district is spending a lot of money on a relative basis for that district, and everybody's got an opinion and so everybody pays attention. If it doesn't go well, if people lose confidence, there is a lot of noise and the superintendent is the one who is the target for the noise or the lightning rod. What often happens is if people have other issues with the superintendent, then they use this as a way to create leverage for their other agendas. The potential here to be a target is high if things don't go well.

The superintendents have very big jobs with a lot to do. They were hired to manage and run these districts and that's more than a full-time job in itself. Now to take on basically another full-time job of developing a process, and then implementing a process to design and build a school, is a huge undertaking. For which we've said earlier, they really don't have a lot of background or training. It's really a high-risk proposition. Certainly there have been a lot of examples of people having situations that occurred that had an impact on their career.

Here Is What's at Stake
for the District...

Susan: I would imagine though, if they get it right, it's a huge feather in their cap.

Kent: It is a huge feather in their cap, and it's actually larger than that. It's a huge feather in the cap of the district. Everybody wins with this. The community wins because the students win and the outcomes improve. One thing that has been shown is when we improve these learning environments, in particular with 21st century learning environments, the co-curricular activities are better delivered. And when they're better delivered, learning outcomes improve. Better learning outcomes benefit the whole community.

Additionally, the district has a reputation. And if you view the reputation as a brand, things that improve the reputation of the district, or tarnish it, impact the brand of the district. Long after the school is built, the reputation of the district lingers and the building endures. There isn't a board of education member that doesn't want a great reputation for its district, and there isn't a community member that doesn't want to be very proud of his or her school district. When it's done well, everybody has a lot of pride, the students benefit, and the district's brand actually improves, which may attract more students.

If it's in certain locations, the district may attract more students. Certainly that successful outcome in the form of the building will be an icon for that community for many years to come, long after the board's gone, long after the superintendent's moved on. Whether successfully done or poorly done, the impact will last for decades.

Susan: Clearly, more than just their job is at stake here, as you pointed out the brand of the district is at stake. I love that. School districts do have brands they need to protect. When a community has a great school district, that often leads to higher property values. There's a lot that goes into this, way beyond just the increased and improved learning, which of course, is critical too.

Kent: Higher property values are another collateral effect. I think that one of the really important things is oftentimes we'll go into a community that maybe 15 years ago had a bond referendum. Whether it passed or failed, people remember all the good things or bad things that were done. Fifteen years is a long time and the board has often turned over and usually they're on either their second or third superintendent. Communities have long memories and so a project that went well, or an experience that went poorly, is long remembered into the future. The organizational memory of the community is very long.

A 21st Century Building Paradigm for a 21st Century School...

Susan: I can imagine. Let's talk about some of the mistakes you've seen or heard that superintendents make when they're trying to manifest this 21st century school with all the different people in play and all the politics behind it. What are some of the mistakes you've seen or heard about?

Kent: One of the biggest challenges of delivering a 21st century learning environment is not embracing some of the 21st century processes that are out there. If we go back to 19th century processes, we're probably going to have a low probability of success. Using an old paradigm of, "Okay, what I need to do is just step back and engage someone to design a building and then go build the building with the lowest bidder."

There are much more collaborative models in place, which bring together teams very early on to help districts develop concepts and provide better outcomes. Through Integrated Project Delivery, this approach is really modeled after manufacturing processes developed in the 1980s that are now called "Lean Manufacturing." It began as "Lean Construction" and is now better known as Integrated Project Delivery. This process starts in the design phase, by assembling the design and construction expertise at the same time, to collaborate using each of their strengths. The design teams' strength is knowing how to design the 21st century building. The construction teams' strength is how to budget, and analyze the constructability options and manage the schedule throughout the process. That's the 21st century paradigm, not the old paradigm. Now the challenge is that a lot of the superintendents don't necessarily know this is an option. The design and construction of the building is a piece of the process, but who can help me cut through the fog of this process and develop the new process? That's really avoiding the "trap" of the old paradigm and adapting to the 21st century model.

Start with the End in Mind...

Susan: I like it, a 21st century building paradigm for a 21st century school. Any mistakes you see, Kent, that you can highlight for the superintendents to avoid?

Kent: The superintendents don't always start with determining an end in mind of what they want and how they might bring everyone along to achieve the best outcome for the students. Sometimes I've seen them get ahead of their boards. They want to move too fast for their staff, or aren't aligned with the community, or not be able to manage the dynamics of competing priorities of multiple stakeholders. All of these can impact, or lessen, their probability of success. With every detractor, or stakeholder that doesn't buy into a process, up go the probabilities of a less than optimum outcome.

Susan: You're saying just like the learning environments the school is creating have changed, the processes they use to create those environments have changed as well.

Kent: Yes, and that isn't just in education, by the way. These processes are very typical out in the marketplace. In workplace environments, there are multiple teams put together to undertake challenges. Rarely are things done in "silos." Whether it's the manufacturing of a car, the design and construction of a building project, or developing curricular materials. There is a need for collaboration among multiple stakeholders to achieve the best outcomes in the process. The approach of assembling a high-performing team, with a diversity of strengths and different perspectives, has shown to get much better outcomes.

Susan: Earlier you pointed out that actual learning environments have changed. You had the old school with the teachers in their own classroom, and now it's much more open and collaborative. I think it's interesting that the old process would have seen the superintendent off in a conference room with a few key players trying to figure this all out, and you're suggesting they tap into a whole new approach to creating this 21st century school. Just as the learning model has evolved, so too has the process to manage building a new school changed.

Kent: Education is now discovering the power of this approach. There's been a lot of research on this concept of Integrated Project Delivery and the research shows it provides lower costs, more satisfied outcomes and faster delivery. That isn't just on the delivery of school projects, that's on all kinds of projects. This idea of collaboration makes sense because those are the benefits that have been demonstrated.

How to Avoid Running Ahead
of the Board...

Susan: You mentioned running ahead of the board of directors. What do you mean by that?

Kent: Sometimes a superintendent moves too quickly down the path of saying, "Here's what I think the solution is for the district," as opposed to saying, "Here's what I think the options are," and letting the board collaborate with the superintendent and key staff on understanding the options. I think in order to have successful outcomes and unity of direction, the superintendent and the board need to see what all the options are, and from this, develop the priorities together of how they want to move forward. In other words, if there are 16 things they want to be done, the "wish list," and the board and the district determine their financial capacity can really only accommodate 11 of them, it's necessary for everyone to align around those 11 through working together to define the top 11 choices.

What's also necessary is for the board to be able to address the staff and the community as to why those 11. In other words, "Here are the 11 we picked, here's why we picked them, here's why it makes sense and here's why we couldn't pick the other 5 (because we can't afford them). In particular, here's why these are the highest priorities." That really should be a collaborative process with the board in order to align the board and superintendent so the district moves forward.

It should be collaborative because the board has to answer to the community, because they are the community representatives, and they need to be deliberate about coming up with solutions. Without that vetting of options and developing aligned priorities, it's really hard to justify why you may not have chosen some of the other options. As I mentioned earlier, we have a lot of competing priorities and we have multiple diverse shareholders. A diverse shareholder may have wanted one or five options that weren't chosen to be part of the program. They need to know why the choices were made, and so does the rest of the community.

Susan: Goodness. These superintendents have to be visionaries. They have to be psychologists. They have to be salesmen...

Kent: "Quarterbacks." They have a team around them that has to be aligned and execute in unison. They have a tough position, and through all this they have to deal with all the "noise" of running the district's day-to-day operations.

Susan: Right, this is on top of an already busy day. Earlier you mentioned the possibility of having to get a bond referendum passed. How does this fit into the process?

Kent: If the district determines it needs more resources, and it has the capacity, one option would be to solicit the community to pass a bond referendum, in order to provide the district the expansion of its financial capacity. A bond referendum requires a significant amount of effort and a very deliberate strategy on how to unfold and explain this to the community, so they will vote for either maintaining a tax or increasing a tax, in support of the bond referendum. It's a very intricate process in and of itself that requires financial skills, marketing skills and legal skills, all orchestrated in unison. How you want to begin to think about that should be at the earliest time as you explore options in order to understand and develop an appropriate strategy for a successful referendum.

Susan: Right, so the superintendents need to think about financial and legal considerations, too.

Kent: Financial, legal and how they will sell the community leaders.

21st Century Thinking...

Susan: I have a whole new level of respect for superintendents and what they have to deal with. The superintendents have a lot on their plates. I want to ask you, we pointed out the big challenges these superintendents face. What are you recommending they do? What is the best way for them to approach building this 21st century school and remain in a favorable position?

Kent: This is a daunting process for sure. It's a high-risk process and it has long-term consequences. Given all of that, we really want to help superintendents deliberately think about how they should approach this. The whole purpose of this book has been to help superintendents develop a way to think about this process that's different from how they might have thought before. In other words, maybe call it 21st century thinking, so 21st century thinking is really about how do we collaborate for better outcomes? Just like 21st century learning.

We want to flesh out an approach or a roadmap of how the superintendent can begin, develop, and end this process, that will lead to a successful outcome. That's really where we start. We try to discover what they are faced with, to pose questions, and then from those questions, to begin to develop a roadmap from concept to completion, with the idea of doing that in a very collaborative way, in order to build support. The superintendent doesn't need to feel like they are there alone. One might say it has the potential to be a very lonely place, a lonely road, and if you think that you've got to travel that lonely road by yourself, you don't.

There are others who understand what they're facing and can help them. The last thing they want to be is the superintendent who goes down the lonely road alone, because that likely leads to a very lonely place. The lonely place is not a good place to be. Our job is to come in and help them develop a collaborative process that will help them formulate how they can begin to unfold this in their district.

How to Tap into Best Practices and Have Peace of Mind...

Susan: Right. A lot of times, they don't know what they don't know at this stage.

Kent: They may not know how to begin to think about this subject or the process. We like to say we help them begin to shift how they should think, entirely and comprehensively, about this subject, and really come up with a collaborative process. The entire district and community will feel great when it's finished. That's the "end in mind" and so it's always hard when a superintendent is beginning to think about this process, what does the end in mind look like? We want to help them visualize what the end in mind might look like, and what might be a pathway, or possible pathways, to get to that end in mind.

Susan: This superintendent you're describing, this may be the project they've undertaken or it's been 20 years, whereas you guys do this every day. Is that right?

Kent: That's exactly right. It's important that a superintendent knows there can be an advocate in their camp, and that's who we are. We're advocates from concept to completion. Not only is it important to know, it's also important to know we do this across multiple districts, so oftentimes that breadth of experience provides reassurance to a superintendent, to a board, and the community. They like to know what other districts are doing, sort of a "best practices." It helps provide peace of mind. For someone to be able to say, "Well, you know, here's how they're doing it in this district or that district."

And while those are different approaches tailored to that community, it still gives them ideas on a best practices approach, and ideas they might consider for their district, to think about how they want to accomplish their outcomes.

Also, the idea of someone assisting to develop a collaborative process for their district, through learning about what is being done in other districts. The "wins" or things to avoid, gives them some reassurance.

Susan: Right, because if you have some insight on what other districts have tried that didn't work well, this particular school district could benefit from that knowledge, and not have to make that same mistake, or go down that same path. That's priceless information to have.

Kent: That's exactly right. Why make the mistakes to learn from the mistakes? In other words, if the successes are there, why not model the successes and learn from others?

Susan: Right, absolutely, every project is unique. They all have unique challenges and unique goals. There's no cookie cutter approach here, but there are probably some things that work and things you've seen not work, and why not tap into that?

Kent: We like to say the "thinking" is still the same, but the approach developed from the thinking is probably what's custom to that community or to that district. How you think about it is kind of the same process. However, the approach that is the outcome from the thinking is very unique and tailored.

Susan: Yeah, very good. This has been very helpful, Kent. I really appreciate you sharing this because I can imagine the daunting task these superintendents are faced with, and as you said, they may think the burden is all on them, and that doesn't have to be the case. In today's more collaborative world, that isn't the case any longer.

Kent: That's right. They don't have to feel alone. There are partners to help them think this through to obtain a successful outcome.

Here's Exactly How to Get Help with Building Your 21st Century School...

Susan: Where can someone go if they want to find these partners?

Kent: If they are in Iowa, we recommend they use their own association, the Iowa Association of School Boards, or IASB, which has developed a process called ICAT. ICAT stands for the Iowa Construction Advocacy Team. If they go to the IASB website at http://www.ia-sb.org/, they can access ICAT program, which we have partnered with IASB to develop. That's a good place for them to begin. We have a brochure that they can download. If not in Iowa, they can access our website, www.estesconstruction.com and our IASB portal to learn more. Either way is a good way to initiate the process. No matter where they are in the project, we can come out, ask some questions and begin to help them discover the right approach for what they are facing.

Susan: Very good. If someone has a specific question for you, how can they reach you?

Kent: Either   through   our   website www.estesconstruction.com or they can call me. My phone number is 563-322-7301.

Susan: On behalf of the superintendents and the school boards, thank you so much, Kent, for taking the time and sharing this information with us today. I can't imagine schools, and learning for the students, won't be better for it.

Kent: You're   welcome,   Susan.   It's   been   my pleasure. We are passionate about helping others be a success.

Here Is How to Leave a Legacy
by Building a School for Tomorrow

You already know building a school for the 21st century is a daunting project, which requires the successful collaboration of many stakeholders. The hardest part is developing an effective process to deal with the competing priorities of running a district, and that's where we come in. We help superintendents just like you create a 21st century process road map, which will lead to a successful outcome for the district.

Here are three steps to start your success.

Step 1: Visit the Iowa Association of School Boards, IASB, website: http://www.ia.-sb.org/ and access the Iowa Construction Advocacy Team (ICAT) program or contact kent@estesconstruction.com

Step 2: We will schedule a visit to learn about what you want to accomplish.

Step 3: Develop an approach or process that "fits" your district.

Most superintendents think they have to manage the school building process on their own and that's just not the case in this 21st century world.

Now you can get your 21st century school built and leave the legacy for your district you have always wanted. If you'd like us to help, just send an email to: kent@estesconstruction.com

About the Author

Education
- Bachelor in Business Administration, Coe College
- Bachelor in Economics, Coe College
- George F. Baker National Business Scholar

Board of Directors Service - Current
- Per Mar Security Company
- Palmer Chiropractic College
  – Board of Trustees
- Skip-A-Long Daycare
  – Board of Directors
- US Bank, Quad City Area
  – Board of Directors
- Quad City AGC
  – Chairman
- Scott County Family YMCA
  – Vice Chairman
- St. Ambrose College of Business Advisory Committee

Leadership Experience
- Past Chairman
  – Master Builders of Iowa
- Past Chairman
  – River Music Experience
- Past Senior Warden
  – St. Peter's Episcopal Church
- Past Chairman
  – Davenport Chamber of Commerce
- Past President
  – Outing Club
- Past President
  – Rejuvenate Davenport

Past Board of Directors Service:
- Scott County YMCA
- Crow Valley Golf Club
- DavenportOne
- Genesis Health System Foundation
- Rejuvenate Davenport
- Coe College – Board of Trustees